BAOFENG RADIO MASTERY

Simplified Strategies for Seamless Off-Grid and Emergency Communication

ARKGUIDE PUBLISHING

Baofeng Radio Mastery

Simplified Strategies for Seamless Off-Grid and Emergency Communication

Arkguide publishing

Copyright

© 2024 Arkguide publishing

All rights reserved.

No part of this book may be reproduced, distributed, or transmitted in any form or by any means, including photocopying, recording, or other electronic or mechanical methods, without the prior written permission of the author, except in the case of brief quotations embodied in critical reviews and certain other noncommercial uses permitted by copyright law.

Disclaimer

The information contained in this book is for general informational purposes only. The author, Arkguide publishing assumes no responsibility for errors or omissions in the contents of the book. While the author has made every effort to ensure that the information in this book was correct at the time of publication, the author does not warrant the accuracy, completeness, or usefulness of any information provided.

The reader should not rely solely on the information within this book for their specific needs and should seek professional advice where appropriate. The author disclaims any liability, loss, or risk, personal or otherwise, which is incurred as a consequence, directly or indirectly, of the use and application of any of the contents of this book.

Introduction	5
Chapter 1: Understanding Baofeng Radios	**9**
A Journey into the World of Baofeng	9
Memory Channels	**26**
Chapter 2: Programming Your Baofeng Radio	**31**
Unlocking the Full Potential of Your Baofeng Radio	31
Chapter 3: Advanced Communication Techniques	**45**
Taking Your Baofeng Radio Skills to the Next Level	45
Chapter 4: Emergency Communication Strategies	**62**
The Crucial Role of Communication in Emergencies	62
Chapter 5: Off-Grid Communication	**75**
Embracing the Freedom of Off-Grid Living	75
Chapter 6: Troubleshooting and Maintenance	**89**
Keeping Your Baofeng Radio in Top Shape	89
Chapter 7: Practical Applications of Your Baofeng Radio	**103**
Everyday Uses and Specialized Scenarios	103
Chapter 8: The Future of Baofeng Radios	**116**
Embracing Tomorrow's Technology Today	116

Introduction

Hello, fellow adventurer!

Welcome to Baofeng Radio Mastery: Simplified Strategies for Seamless Off-Grid and Emergency Communication. Over the past decade, I've spent countless hours exploring the capabilities of these remarkable devices, and now I'm here to share my knowledge with you.

I remember vividly the first time I held a Baofeng radio in my hands. It was during a family camping trip, deep in the mountains, far from any cell phone signal. The sense of security and connection it provided was unparalleled.

Since then, I've used Baofeng radios in a variety of situations— from coordinating off-road adventures with friends to staying connected during power outages and emergencies.

These experiences have shown me just how indispensable a well-understood and properly utilized radio can be.

Why should you care about mastering your Baofeng radio? Picture this: You're on a solo hike, enjoying the serenity of nature, when suddenly you find yourself off the trail with no signal on your phone. Or imagine a severe storm has knocked out power and cell service in your area, leaving you cut off from loved ones. In these moments, having a reliable means of communication is not just a convenience—it's a lifeline.

This book is designed to equip you with everything you need to know to confidently use your Baofeng radio in any situation. We'll cover the basics of setup and operation, delve into advanced communication techniques, and explore practical applications that make these radios so versatile.

You'll learn how to effectively communicate during emergencies, set up a reliable off-grid communication network, and troubleshoot common issues.

But this isn't just a technical manual. It's a companion guide filled with real-life stories, practical tips, and insights gained from years of hands-on experience. I want you to feel as if we're having a conversation, with me sharing the lessons I've learned along the way.

My goal is to make this journey enjoyable and engaging, so you not only understand how to use your Baofeng radio but also appreciate the incredible potential it offers.

Whether you're a seasoned radio operator or a complete beginner, this book has something for you.

We'll start with the fundamentals and gradually move into more advanced topics, ensuring you have a solid foundation before tackling the more complex aspects of radio communication.

By the end of this book, you'll have the confidence and knowledge to use your Baofeng radio effectively, no matter where you are or what challenges you face.

So, are you ready to unlock the full potential of your Baofeng radio? Let's get started on this exciting journey together. With each page, you'll gain new skills and insights that will make you a master of seamless off-grid and emergency communication.

Chapter 1: Understanding Baofeng Radios

A Journey into the World of Baofeng

These little devices might seem daunting at first, but trust me, by the end of this book, you'll feel like a pro. In this chapter, we'll start with the basics—understanding what Baofeng radios are, their key features, and why they've become such an essential tool for so many people.

A Brief History of Baofeng Radios

First, a little history. Baofeng radios are manufactured by Baofeng Tech, a company based in China.

They entered the market with a mission: to create affordable, high-quality communication tools accessible to everyone. And they've certainly succeeded.
Today, Baofeng radios are used by hobbyists, preppers, outdoor enthusiasts, and professionals worldwide.

Why Baofeng? It's a question I get a lot. The answer is simple—versatility and affordability. Baofeng radios offer a wide range of features at a price point that makes them accessible to almost anyone. They're not just for tech geeks or survivalists; they're for anyone who values reliable communication.

Key Features and Models

Let's talk about what makes Baofeng radios special. Here are some key features that have made them so popular:

1. Dual Band Functionality: Most Baofeng radios operate on both VHF (Very High Frequency) and UHF (Ultra High Frequency) bands.
This dual-band capability gives you more flexibility in finding clear channels and communicating over different distances and terrains.

2. Programmable Channels: Baofeng radios come with programmable channels, allowing you to customize your device to meet your specific needs. Whether you're setting it up for local use or preparing for an emergency, you can program the channels that are most relevant to you.

3. Affordable: One of the biggest selling points of Baofeng radios is their affordability. You get a lot of features for a fraction of the cost of other brands. This makes them an excellent choice for beginners or anyone on a budget.

4. Durability: Despite their low price, Baofeng radios are surprisingly durable. They can withstand rough handling, making them suitable for outdoor activities and harsh environments.

5. Wide Range of Accessories: There's a vast array of accessories available for Baofeng radios. From high-gain antennas to extended batteries, you can customize your radio to fit your exact needs.

Now, let's look at some popular models:

- Baofeng UV-5R: This is probably the most popular model and for good reason. It's affordable, versatile, and packed with features. If you're just starting out, the UV-5R is a great choice.

- Baofeng BF-F8HP: This is an upgraded version of the UV-5R, offering more power and better performance.
It's ideal if you need a bit more range and durability.

- Baofeng UV-82: Known for its rugged design and improved audio quality, the UV-82 is perfect for outdoor enthusiasts and professionals who need reliable communication in tough conditions.

Basics of Radio Communication

Before we get into the nitty-gritty of operating your Baofeng radio, it's important to understand some basic concepts of radio communication.

Frequency Bands

Radio communication relies on transmitting and receiving signals over different frequency bands. Here's a quick rundown:

- VHF (Very High Frequency): VHF operates in the range of 30 MHz to 300 MHz. It's great for long-distance communication in open areas but struggles with obstacles like buildings and hills.

- UHF (Ultra High Frequency): UHF operates in the range of 300 MHz to 3

GHz. It's better at penetrating obstacles, making it ideal for urban environments.

Your Baofeng radio's ability to switch between these bands gives you the flexibility to communicate effectively in a variety of situations.

Channels and Frequencies

Channels are essentially preset frequencies that your radio can tune into. When you program a channel, you're setting your radio to a specific frequency that you want to use for communication.

This allows you to easily switch between different communication lines without manually adjusting the frequency each time.

Licensing

In many countries, including the United States, operating on certain frequencies requires a license. This is especially true for frequencies in the amateur (ham) radio bands.

The good news is that getting a license is relatively straightforward and opens up a whole new world of communication possibilities. We'll discuss more about licensing in later chapters.

Squelch and Tone Codes

- Squelch: Squelch is a feature that mutes the audio of your radio when the signal is too weak to be heard clearly. This helps reduce background noise and makes communication more pleasant.

- CTCSS and DCS: Continuous Tone-Coded Squelch System (CTCSS) and Digital-Coded Squelch (DCS) are systems used to reduce interference by adding a tone to your transmission. Only radios set to the same tone can communicate with each other, which helps keep your conversations private.

Unboxing and Initial Setup

Now that you have a basic understanding of how Baofeng radios work, let's get hands-on. When you first unbox your Baofeng radio, you'll find several key components:

1. The Radio: The main device itself.

2. Antenna: This attaches to the radio to allow it to transmit and receive signals.

3. Battery: The power source for your radio.

4. Charger: To keep your battery full and ready to go.

5. Belt Clip: Handy for attaching the radio to your belt or bag.

6. Earpiece: Optional, but useful for hands-free operation.

7. User Manual: A basic guide to your radio's features.

Charging Your Radio

Before you start using your radio, you'll need to charge the battery. Here's a simple step-by-step guide:

1. Insert the Battery: Slide the battery into the back of the radio until it clicks into place.

2. Connect the Charger: Plug the charger into a power outlet and connect it to the radio.

3. Charging Indicator: Most chargers have an LED indicator. It will usually show red while charging and green when fully charged.

4. Full Charge: It's best to let your radio charge fully before its first use. This might take a few hours, so be patient.

Assembling Your Radio

Once the battery is charged, it's time to assemble your radio:

1. Attach the Antenna: Screw the antenna onto the top of the radio. Make sure it's snug but don't over-tighten it.

2. Belt Clip: If you plan to carry the radio on your belt, attach the belt clip to the back of the radio using the provided screws.

3. Earpiece: If you're using the earpiece, plug it into the appropriate jack on the side of the radio.

Powering Up

Now for the exciting part—turning on your radio for the first time:

1. Power Button: Hold down the power button until the radio turns on. You'll hear a welcome message and see the display light up.

2. Volume Control: Adjust the volume to a comfortable level using the knob on the top of the radio.

Basic Operations: Getting to Know Your Radio

Let's familiarize ourselves with the basic functions of your Baofeng radio. Don't worry if it feels a bit overwhelming at first—we'll take it one step at a time.

- Channel Selector: Use the knob on the top to switch between different channels. Each click represents a different frequency.

- Push-to-Talk (PTT) Button: This is the key to communication. Hold it down to talk and release it to listen.

- Menu Button: Access the radio's menu for advanced settings. We'll dive into this in more detail later.

- LCD Display: Shows important information like the current channel, battery level, and signal strength.

- Keypad: Use the buttons to navigate menus and input frequencies manually.

Making Your First Call

Let's make your first call. Here's how to do it:

1. Select a Channel: Use the channel selector to choose an open channel. For now, stick to the pre-programmed channels.

2. Push-to-Talk: Hold down the PTT button and speak clearly into the microphone. Start with something simple like, "Testing, 1, 2, 3."

3. Release and Listen: Release the PTT button to listen for a response. If someone is on the other end, you'll hear their reply.

Safety and Etiquette

Before we go further, let's talk about some basic safety and etiquette rules:

- Respect Privacy: Avoid transmitting on channels that are in use, and never listen in on private conversations.

- Legal Use: Ensure you're following local regulations regarding frequency use and licensing.

- Clear Communication: Speak clearly and concisely. Avoid unnecessary chatter, especially on emergency channels.

- Emergency Protocols: Familiarize yourself with emergency protocols and the specific frequencies designated for emergency use.

Memory Channels

Your Baofeng radio can store multiple frequencies in its memory channels.
This allows you to quickly switch between commonly used frequencies without having to manually input them each time. Here's how to program a memory channel:

1. Enter Frequency Mode: Press the "VFO/MR" button to switch to frequency mode.

2. Input Frequency: Use the keypad to enter the desired frequency.

3. Access Menu: Press the "MENU" button and navigate to the "MEM-CH" setting.
4. Select Channel: Choose an empty memory channel and press "MENU" to save the frequency.

Repeat these steps for any additional frequencies you want to program. Having your most-used frequencies saved can save you a lot of time and hassle.

Scanning Channels

Your Baofeng radio has a scan feature that allows you to search for active frequencies. This is useful for finding open channels or monitoring multiple frequencies at once. Here's how to use it:

1. Enter Scan Mode: Press and hold the "SCAN" button.

2. Start Scanning: The radio will begin scanning through the channels. It will stop when it detects an active frequency.

3. Exit Scan Mode: Press any button to stop the scan.

Scanning is a great way to discover new frequencies and monitor activity in your area.

Troubleshooting Common Issues

Starting out, you might encounter a few hiccups. Here are some common issues and how to solve them:

- No Power: Ensure the battery is charged and properly installed. Double-check the battery contacts for any dirt or corrosion.

- Poor Signal: Check the antenna and make sure it's securely attached. Try moving to a different location to improve signal strength.

- No Sound: Adjust the volume and ensure the radio is not on mute. Check the earpiece connection if you're using one.

- Can't Transmit: Verify you're on the correct channel and that it's not already in use. Double-check your CTCSS/DCS settings.

Troubleshooting can be frustrating, but it's part of the learning process. Don't be afraid to consult the user manual or seek help from online communities if you run into issues.

You've now taken your first steps into the world of Baofeng radios. You've learned how to set up your radio, make a call, and explore some of its advanced features. Great job!

Next we'll dive into programming your radio, both manually and using software. This is where the real fun begins. You'll learn how to customize your radio to fit your specific needs and preferences, making it an even more powerful tool.

Spend some time getting comfortable with the basic operations of your radio. The more you use it, the more intuitive it will become.

Chapter 2: Programming Your Baofeng Radio

Unlocking the Full Potential of Your Baofeng Radio

Now that you're familiar with the basics and have set up your Baofeng radio, it's time to dive into programming.

This chapter will guide you through both manual programming and using software like CHIRP. Programming your radio will allow you to customize it for your specific needs and make it much more versatile.

Manual Programming: Step-by-Step Guide

Manual programming can seem daunting at first, but don't worry. I'll walk you through it step by step.

By the end of this section, you'll be able to program your Baofeng radio with confidence.

Step 1: Enter Frequency Mode

1. Turn On the Radio: Hold down the power button until your radio turns on.

2. Frequency Mode: Press the "VFO/MR" button to switch to Frequency Mode. You should see "VFO" on the display.

Step 2: Select the Frequency

1. Enter the Frequency: Use the keypad to enter the frequency you want to program. For example, if you want to set it to 146.520 MHz, just type 146520.

Step 3: Set CTCSS/DCS Codes

1. Menu Access: Press the "MENU" button to access the menu.

2. CTCSS/DCS Settings: Use the arrow keys to navigate to the CTCSS (Menu 13) or DCS (Menu 10) settings.

3. Set Codes: Press "MENU" to select, use the arrow keys to choose your desired code, and press "MENU" again to confirm. Finally, press "EXIT" to exit the menu.

Step 4: Save to a Channel

1. Menu Access: Press "MENU" again to access the menu.

2. Save Channel: Navigate to "MEM-CH" (Menu 27).

3. Choose Channel: Use the arrow keys to select an empty channel number (e.g., CH-01).

4. Save: Press "MENU" to save and "EXIT" to finish.

Congratulations! You've manually programmed a frequency into your Baofeng radio. Repeat these steps for any additional frequencies you want to program.

Using CHIRP Software: Simplifying the Process

While manual programming is handy, using software like CHIRP can simplify the process, especially when you need to program multiple frequencies or settings. Here's how to do it:

Step 1: Download and Install CHIRP

1. Download CHIRP: Go to the [CHIRP website](https://chirp.danplanet.com/projects/chirp/wiki/Download) and download the software compatible with your operating system.

2. Install CHIRP: Follow the installation instructions specific to your OS.

Step 2: Connect Your Radio

1. Cable Connection: Connect your Baofeng radio to your computer using a programming cable. Make sure the radio is turned off before connecting.

2. Turn On the Radio: Once connected, turn on your radio.

Step 3: Configure CHIRP

1. Open CHIRP: Launch the CHIRP software on your computer.

2. Download From Radio: Go to "Radio" in the menu and select "Download From Radio." Follow the prompts to select your radio model and COM port.

3. Read Data: Click "OK" to read the data from your radio. This may take a few moments.

Step 4: Program Frequencies

1. Frequency List: You'll see a spreadsheet-like interface where you can enter frequencies, names, tones, and other settings.

2. Add Frequencies: Enter the frequencies and settings you want to program. You can find frequency lists online for your area or specific needs.

3. Save Changes: Once you've entered all the necessary information, go to "File" and select "Save" to save your work.

Step 5: Upload to Radio

1. Upload Data: Go to "Radio" in the menu and select "Upload To Radio."

2. Confirm Upload: Follow the prompts to confirm the upload to your radio. This may take a few moments.

3. Finish: Once the upload is complete, you can disconnect your radio and test the newly programmed frequencies.

Using CHIRP can save you a lot of time and makes it easy to manage multiple channels and settings.

Advanced Programming Techniques

Now that you've mastered the basics of programming, let's explore some advanced techniques that can enhance your radio's functionality.

Using CTCSS and DCS Codes

CTCSS (Continuous Tone-Coded Squelch System) and DCS (Digital-Coded Squelch) are systems used to reduce interference and ensure privacy. Here's how to set them up:

1. Menu Access: Press the "MENU" button to access the menu.

2. Navigate to Settings: Use the arrow keys to navigate to CTCSS (Menu 13) or DCS (Menu 10).

3. Set Codes: Select the desired code and press "MENU" to confirm. This will filter out unwanted transmissions on the same frequency.

Accessing Repeaters

Repeaters are stations that receive a signal and retransmit it at a higher power, extending the range of your communication. Here's how to use them:

1. Frequency Split: Set the radio to the repeater's receive frequency.

2. Offset Frequency: Set the offset frequency, which is the difference between the repeater's receive and transmit frequencies.

3. Activate Offset: Access the menu and set the "OFFSET" (Menu 26) to the desired value.

Using repeaters can significantly increase your communication range, making it easier to stay in touch over long distances.

Troubleshooting Programming Issues

Sometimes things don't go as planned. Here are some common programming issues and how to fix them:

- No Communication: Ensure both radios are on the same frequency and using the same CTCSS/DCS codes.

- Invalid Frequency: Double-check that the frequency is within the allowed range for your radio model.

- Software Errors: Ensure you're using the correct driver for your programming cable and that the COM port settings match.

Practical Tips for Effective Programming

Here are some additional tips to help you program your Baofeng radio more effectively:

Organizing Your Frequencies

1. Group Similar Frequencies: Organize your frequencies into groups based on their use (e.g., emergency, local, hobby). This makes it easier to find the right channel quickly.

2. Use Clear Labels: When using software like CHIRP, label each channel with a clear name. This helps you identify channels at a glance.

Backup Your Settings

1. Save Configurations: Regularly save your radio's configurations using CHIRP. This ensures you can quickly restore your settings if needed.

2. Print a List: Keep a printed list of your programmed frequencies and settings as a backup.

Practice Makes Perfect

1. Regular Practice: Regularly practice programming and using your radio. The more you use it, the more comfortable you'll become.

2. Stay Updated: Keep up with updates and new programming techniques. Join online forums and communities to learn from other users.

You've now learned how to program your Baofeng radio both manually and using software.

This skill is crucial for customizing your radio to suit your specific needs, making it a versatile tool for various situations.

Remember, every expert was once a beginner. Keep practicing, stay curious, and don't hesitate to experiment with new settings and configurations. Stay tuned for more exciting adventures with your Baofeng radio!

Chapter 3: Advanced Communication Techniques

Taking Your Baofeng Radio Skills to the Next Level

Now that you've mastered the basics of operating and programming your Baofeng radio, it's time to dive into advanced communication techniques. In this chapter, we'll cover how to use repeaters, optimize your signal strength, and ensure your communications are secure.

These skills will help you get the most out of your Baofeng radio, whether you're in an urban environment or out in the wilderness.

Using Repeaters

Repeaters are a powerful tool that can significantly extend the range of your communications. By receiving and retransmitting signals, repeaters allow you to communicate over much greater distances than you could with just your handheld radio.

What is a Repeater?

A repeater is a radio system that receives a signal on one frequency and retransmits it on another. This process boosts the signal, allowing it to cover a much larger area. Repeaters are typically located on high buildings, towers, or mountaintops to maximize their range.

Finding a Repeater

To use a repeater, you first need to find one in your area. Here are some tips on how to locate repeaters:

1. Online Databases: Websites like [RepeaterBook](https://www.repeaterbook.com/) and [ARRL Repeater Directory](http://www.arrl.org/repeater-directory) list repeaters by location.

2. Local Radio Clubs: Joining a local amateur radio club can provide valuable information about repeaters in your area.

3. Scanning: Use your radio's scan feature to search for active repeaters.

Programming a Repeater into Your Radio

Once you've identified a repeater, you need to program it into your radio. Here's how:

1. Enter Frequency Mode: Press the "VFO/MR" button to switch to Frequency Mode.

2. Input Receive Frequency: Enter the repeater's receive frequency using the keypad.

3. Set the Offset: Access the menu and navigate to the "OFFSET" setting (Menu 26). Set the offset frequency to the repeater's specified offset.

4. Activate Offset: Navigate to the "SFT-D" setting (Menu 25) and set it to "+" or "-" depending on the repeater's offset direction.

5. Set CTCSS/DCS Code: If the repeater requires a tone, navigate to the

"T-CTCS" (Menu 13) or "T-DCS" (Menu 10) setting and input the required code.

6. Save to Memory: Save the repeater settings to a memory channel using the "MEM-CH" setting (Menu 27).

Optimizing Signal Strength and Range

Optimizing your signal strength is crucial for ensuring clear and reliable communication. Here are some tips to help you get the best performance from your Baofeng radio:

Antenna Placement and Upgrades

1. Antenna Placement: Always hold your radio vertically with the antenna pointing upwards. This position helps maximize signal transmission and reception.

2. High-Gain Antennas: Upgrading to a high-gain antenna can significantly improve your radio's range. Antennas like the Nagoya NA-771 are popular choices.

3. External Antennas: If you're stationary, consider using an external antenna mounted on a higher location, like a roof or a mast.

Power Settings

1. Adjusting Power Levels: Your Baofeng radio allows you to adjust the power output. Use the high power setting (typically 5 watts) for long-range communication and the low power setting (1 watt) for short-range to conserve battery life.

2. Battery Maintenance: Keep your batteries charged and carry spares if you

plan to use your radio for extended periods. Extended-life batteries can also be a great investment.

Location and Environment

1. Line of Sight: Radio signals travel best in a straight line. Position yourself in open areas with a clear line of sight to your communication partner or repeater.

2. Avoid Obstacles: Buildings, trees, and hills can obstruct radio signals. If possible, move to higher ground or a more open area to improve signal strength.

3. Weather Conditions: Weather can affect radio signals. Be aware that rain, snow, and fog might reduce your effective communication range.

Enhancing Privacy and Security

Ensuring the privacy and security of your communications is essential, especially if you're discussing sensitive information. Here are some techniques to enhance your radio's privacy and security:

CTCSS and DCS Codes

Using Continuous Tone-Coded Squelch System (CTCSS) and Digital-Coded Squelch (DCS) codes can help reduce interference and ensure that only radios set to the same code can communicate with each other.

1. Setting CTCSS Codes: Access the menu and navigate to the "T-CTCS" setting (Menu 13). Select the desired tone and press "MENU" to confirm.

2. Setting DCS Codes: Navigate to the "T-DCS" setting (Menu 10). Select the

desired code and press "MENU" to confirm.

Encryption

Some advanced radios offer encryption features to further secure your communications. While Baofeng radios don't typically include sophisticated encryption, using less common frequencies and regularly changing CTCSS/DCS codes can add a layer of security.

Frequency Hopping

Manually changing frequencies during your communication can help avoid interception. This method, known as frequency hopping, makes it harder for eavesdroppers to follow your conversation.

1. Pre-Plan Frequencies: Decide on a set of frequencies and the order in which you'll switch between them.

2. Communicate the Plan: Ensure all parties involved know the frequency hopping plan.

3. Regularly Switch: Change frequencies at predetermined intervals to enhance security.

Troubleshooting Advanced Communication Issues

Even with advanced techniques, you might encounter issues. Here's how to troubleshoot some common problems:

Weak Signal or Interference

1. Antenna Check: Ensure the antenna is securely attached and not damaged.

2. Change Location: Move to a higher or more open area to improve signal strength.

3. Adjust Power Levels: Increase the power output if you're experiencing weak signals.

No Communication Through Repeater

1. Check Offset and CTCSS/DCS Codes: Ensure the offset frequency and tone codes are correctly set.

2. Monitor Repeater Status: Verify that the repeater is active and not undergoing maintenance.

Distorted or Unclear Audio

1. Volume and Squelch: Adjust the volume and squelch settings to reduce background noise.

2. Microphone Check: Ensure the microphone is not obstructed or damaged. Speak clearly and directly into it.

Practical Applications of Advanced Techniques

To illustrate how these advanced techniques can be used, let's revisit some of our earlier scenarios and see how they benefit from advanced communication strategies:

Emily the Hiker

When Emily and her hiking group encountered severe weather, using a repeater allowed them to communicate over long distances and coordinate their regrouping more effectively.

By optimizing their signal strength and using high-gain antennas, they maintained clear communication despite the challenging conditions.

John in the Urban Emergency

During the city-wide blackout, John's neighborhood watch group used CTCSS codes to keep their communications private and secure. They also set up a frequency hopping plan to avoid interference from other radio users, ensuring their messages got through without disruption.

David the Off-Roader

David's off-roading team used external antennas mounted on their vehicles to extend their communication range. By utilizing repeaters and adjusting power levels, they stayed in touch even in remote areas with challenging terrain.

Rachel the Event Organizer

At the local fair, Rachel's team used frequency hopping to maintain secure communications and avoid interference from other radio users. This ensured that security updates and logistical information were transmitted clearly and efficiently.

You've now learned how to use advanced communication techniques to get the most out of your Baofeng radio. From using repeaters and optimizing signal strength to enhancing privacy and security, these skills will make you a more proficient and confident radio operator.

Next we'll focus on emergency communication strategies. We'll explore how to set up a reliable communication plan for emergencies, essential frequencies to know, and practical tips for staying connected when it matters most.

Keep practicing these advanced techniques and experimenting with your radio. The more you use and explore, the more adept you'll become.

Chapter 4: Emergency Communication Strategies

The Crucial Role of Communication in Emergencies

Now that you've honed your skills with your Baofeng radio, it's time to focus on one of the most critical applications: emergency communication. When disaster strikes, reliable communication can be the difference between safety and peril.

Why Communication is Critical

Imagine a severe storm hitting your area, knocking out power and cell phone service. You're cut off from news updates, emergency services, and even your family.

In such scenarios, having a Baofeng radio and knowing how to use it effectively can be a lifesaver. Reliable communication allows you to stay informed, coordinate with others, and call for help if needed.

Creating a Reliable Communication Plan

Having a communication plan in place before an emergency occurs is essential. Here's how to create one:

Step 1: Identify Key Contacts

- Family Members: Ensure everyone in your household knows how to use the Baofeng radio and understands the communication plan.

- Neighbors: Establish a network with your neighbors. Knowing you can communicate with them during an emergency can provide additional support and resources.

- Emergency Services: Familiarize yourself with local emergency frequencies and channels. These are often monitored by first responders and can be critical during a disaster.

Step 2: Choose Primary and Backup Frequencies

- Primary Frequency: Select a primary frequency for your group to use during an emergency. Ensure everyone knows this frequency and how to access it.

- Backup Frequencies: Have several backup frequencies in case the primary one is crowded or unusable. Program these into your Baofeng radio ahead of time.

Step 3: Establish Communication Protocols

- Check-In Times: Set regular check-in times to ensure everyone is safe and accounted for. For example, you might decide to check in every hour.

- Emergency Codes: Develop simple codes to convey urgent information quickly. For example, "Code Red" could mean immediate danger, while "Code Green" could indicate all is well.

Step 4: Practice Your Plan

- Drills: Regularly practice your communication plan with all members of your group. This ensures everyone knows what to do and can operate their radios confidently.

- Review and Adjust: Periodically review your plan and make adjustments as needed.

Changes in your group or environment might require updates to your plan.

Essential Emergency Frequencies and Channels

Knowing which frequencies to use during an emergency is crucial. Here are some commonly used emergency frequencies:

- NOAA Weather Radio: Frequencies between 162.400 MHz and 162.550 MHz provide weather updates and alerts.

- MURS (Multi-Use Radio Service): Frequencies like 151.820 MHz and 154.600 MHz are used for local communication and do not require a license.

- FRS/GMRS (Family Radio Service/General Mobile Radio Service): Frequencies in the 462 MHz range are commonly used for local communication. Note that GMRS requires a license.

- Amateur Radio (Ham): Frequencies in the 144-148 MHz (VHF) and 420-450 MHz (UHF) ranges are used by licensed Ham radio operators for emergency communication.

Setting Up Your Baofeng Radio for Emergencies

Let's get your Baofeng radio ready for emergency use. Here's a step-by-step guide:

Step 1: Program Emergency Frequencies

- Manual Programming: Follow the steps outlined in Chapter 3 to manually program emergency frequencies into your radio.

- CHIRP Software: Use CHIRP to program multiple frequencies quickly and efficiently.
Ensure all emergency frequencies are saved and labeled clearly.

Step 2: Save Important Channels

- Priority Channels: Save key frequencies as priority channels for easy access. This ensures you can quickly switch to important frequencies during an emergency.

- Channel Scan: Enable the channel scan feature to monitor multiple frequencies simultaneously. This helps you stay informed about any developments.

Step 3: Prepare Backup Power

- Spare Batteries: Keep spare batteries charged and ready. This ensures your radio remains operational even if power is out for an extended period.

- Alternative Power Sources: Consider investing in solar chargers or hand-crank generators to keep your batteries charged during long outages.

Emergency Communication Protocols

During an emergency, following proper communication protocols is essential. Here are some tips:

1. Clear and Concise Messages

   - Keep your messages short and to the point. This ensures critical information is communicated quickly and effectively.

2. Prioritize Urgent Information

- Prioritize transmitting urgent information, such as safety updates and emergency alerts. This helps ensure everyone receives the most critical information first.

3. Stay Calm and Composed

- In high-stress situations, staying calm can help ensure your message is clear. Take a deep breath before speaking and avoid shouting or rushing your words.

4. Listen Before Transmitting

- Always listen to the channel before transmitting to ensure it is clear. This prevents talking over others and ensures your message is heard.

5. Use Plain Language

- Avoid using codes or jargon unless everyone in your group understands them. Clear, plain language ensures your message is understood by everyone.

Real-Life Applications and Success Stories

To illustrate the importance of these skills, let me share a story about a friend of mine, Ethan. During a severe winter storm, Ethan's neighborhood lost power and all forms of communication.

Using his Baofeng radio, he coordinated with his neighbors to check on elderly residents, share food and resources, and stay informed about weather updates. His preparedness and knowledge of his radio's capabilities made all the difference in ensuring the safety and well-being of his community.

Troubleshooting in Emergencies

Even in emergencies, you might encounter issues. Here's how to troubleshoot common problems:

- No Signal: Check your antenna and ensure it's properly connected. Move to a higher location if possible.

- Interference: Switch to a different frequency or use CTCSS/DCS codes to reduce interference.

- Battery Issues: Swap out batteries or use an alternative power source if your battery is running low.

You've now learned how to use your Baofeng radio for disaster preparedness, ensuring you can stay connected and informed during emergencies. In the next chapter, we'll focus on off-grid communication. We'll cover preparing for off-grid adventures, choosing the right frequencies, and practical tips for effective off-grid communication.

Remember, being prepared is more than just having the right tools—it's knowing how to use them effectively. Keep practicing, stay vigilant, and continue refining your skills.

Chapter 5: Off-Grid Communication

Embracing the Freedom of Off-Grid Living

We've covered a lot of ground so far, and now it's time to dive into one of the most exciting applications of your Baofeng radio: off-grid communication. Whether you're planning a remote camping trip, going off the grid for an extended period, or simply preparing for a situation where conventional communication might fail, mastering off-grid communication is essential.

Preparing for Off-Grid Adventures

Preparation is key to ensuring your off-grid communication setup is reliable and effective. Here's a checklist to get you started:

Equipment Essentials

1. Baofeng Radio: Ensure your radio is fully charged and in good working condition.

2. Spare Batteries: Carry fully charged spare batteries. Consider investing in high-capacity or extended-life batteries.

3. Antenna Upgrades: A high-gain or flexible antenna can significantly improve your range and signal strength.

4. Solar Charger/Hand Crank Generator: These are invaluable for keeping your batteries charged when you're away from conventional power sources.

5. Protective Case: A rugged case can protect your radio from the elements and rough handling.

Planning Your Communication

1. Communication Plan: Establish a clear communication plan with your group. Decide on primary and backup frequencies, check-in times, and emergency protocols.

2. Familiarize Yourself with the Area: Know the terrain and any potential obstacles that might affect your signal. Mountains, dense forests, and valleys can all impact radio communication.

3. Local Frequencies: Research and program local frequencies, including emergency channels and local repeaters, into your radio.

Practicing Skills

1. Regular Drills: Conduct regular communication drills with your group to ensure everyone is comfortable using the radios and understands the communication plan.

2. Signal Testing: Test your signal strength and range in different environments. This will help you understand how far you can communicate and under what conditions.

Choosing the Right Frequencies and Channels

Selecting the right frequencies and channels is crucial for effective off-grid communication. Here's a guide to help you choose:

Common Off-Grid Frequencies

1. MURS (Multi-Use Radio Service): Frequencies like 151.820 MHz and 154.600 MHz are popular for off-grid communication. They do not require a license and are often less crowded.

2. FRS/GMRS (Family Radio Service/General Mobile Radio Service): Frequencies in the 462 MHz range are commonly used for local communication. GMRS requires a license, but it offers better range and more power.

3. Amateur Radio (Ham): Frequencies in the 144-148 MHz (VHF) and 420-450 MHz (UHF) ranges are used by licensed Ham radio operators. These bands provide excellent range and reliability.

Programming Frequencies

Here's a step-by-step guide to programming your chosen frequencies:

1. Enter Frequency Mode: Press the "VFO/MR" button to switch to Frequency Mode.

2. Input Frequency: Use the keypad to enter the desired frequency.

3. Set CTCSS/DCS Codes: If needed, access the menu to set the CTCSS or DCS codes for added privacy and reduced interference.

4. Save to Memory: Save the programmed frequency to a memory channel for easy access.

Using Repeaters

Repeaters can be a game-changer for off-grid communication by extending your range. Here's how to find and use them:

1. Finding Repeaters: Use online resources like [RepeaterBook](https://www.repeaterbook.com/) to locate repeaters in your area.

2. Programming Repeaters: Follow the steps outlined in Chapter 4 to program repeaters into your radio.

3. Testing Repeaters: Test your connection with repeaters before heading off-grid to ensure they're working correctly.

Practical Tips for Effective Off-Grid Communication

Here are some practical tips to ensure your off-grid communication is as effective as possible:

Optimizing Signal Strength

1. Antenna Positioning: Hold your radio vertically with the antenna pointing upwards. Position yourself in open areas with a clear line of sight.

2. Higher Ground: Elevation can significantly improve your signal. Move to higher ground if possible.

3. Avoid Obstacles: Buildings, trees, and other obstructions can weaken your signal. Try to find clear, open spaces for communication.

Battery Management

1. Conserve Power: Use the low power setting for short-range communication to conserve battery life. Switch to high power only when needed.

2. Solar Charging: Use a solar charger during the day to keep your batteries topped up. Position the charger in direct sunlight for maximum efficiency.

3. Hand Crank Generators: These can be a lifesaver in cloudy conditions or at night. A few minutes of cranking can provide enough power for essential communications.

Emergency Protocols

1. Pre-Set Emergency Channels: Program local emergency frequencies into your radio and familiarize yourself with their usage.

2. Emergency Codes: Develop simple emergency codes with your group to quickly convey urgent information.

3. Regular Check-Ins: Establish regular check-in times with your group to ensure everyone's safety and well-being.

Real-Life Scenario: Off-Grid Camping Trip

Let me share a story about a friend, Marie, who went on a week-long camping trip in a remote forest. She and her group relied heavily on their Baofeng radios for communication. Here's how they made it work:

1. Preparation: Marie programmed local frequencies, including nearby repeaters, and established a communication plan with her group.

2. Equipment: They carried spare batteries, solar chargers, and a hand crank generator to ensure they always had power.

3. Signal Testing: Before setting up camp, they tested their radios in various locations to find the best signal spots.

4. Regular Check-Ins: The group had scheduled check-ins every few hours to update each other on their locations and any changes in plans.

5. Emergency Protocols: They had pre-set emergency codes and knew the local emergency frequencies.

During the trip, one of the group members got lost while hiking. Thanks to their preparation and reliable communication, Marie was able to use her Baofeng radio to coordinate a quick and effective search.

The lost hiker was found safe and sound, and the group continued their adventure without further incident.

Troubleshooting Off-Grid Communication Issues

Even with the best preparation, issues can arise. Here's how to troubleshoot common problems:

Weak Signal

1. Antenna Check: Ensure the antenna is securely attached and not damaged.

2. Change Location: Move to a higher or more open area to improve signal strength.

3. Adjust Power Levels: Increase the power output if you're experiencing weak signals.

Battery Issues

1. Conserve Power: Switch to low power mode for short-range communication.

2. Check Connections: Ensure the battery is properly connected and not loose.

3. Use Alternative Power: Utilize solar chargers or hand crank generators to keep your batteries charged.

Interference

1. Change Frequency: Switch to a different frequency if you're experiencing interference.

2. Set CTCSS/DCS Codes: Use these codes to filter out unwanted transmissions.

3. Monitor Environment: Be aware of potential sources of interference, such as electronic devices or power lines, and move away from them.

Congratulations! You've now learned how to effectively use your Baofeng radio for off-grid communication. From preparing your equipment and choosing the right frequencies to practical tips and troubleshooting, you're well-equipped to stay connected in remote locations.

Remember, practice and preparation are key to mastering off-grid communication. Keep testing your skills, refining your setup, and staying informed about new techniques and technologies.

Chapter 6: Troubleshooting and Maintenance

Keeping Your Baofeng Radio in Top Shape

We've covered a lot of ground so far, and now it's time to focus on keeping your Baofeng radio in top shape. Regular maintenance and troubleshooting are essential for ensuring your radio's longevity and reliability.

Common Issues and Their Solutions

Even the most reliable devices can encounter problems. Here are some common issues you might face with your Baofeng radio and how to troubleshoot them:

1: No Power

1. Check the Battery
    - Ensure the battery is properly connected and fully charged. Try using a spare battery if available.

2. Inspect the Contacts
    - Check the battery contacts for dirt or corrosion. Clean them gently with a dry cloth if needed.

3. Reset the Radio
    - Turn off the radio, remove the battery, wait a few seconds, then reinsert the battery and turn the radio back on.

2: Poor Signal Quality

1. Antenna Check
    - Ensure the antenna is securely attached. A loose or damaged antenna can affect signal quality.

2. Location Matters

- Move to a higher location or an open area to improve signal strength. Buildings, trees, and other obstacles can interfere with signals.

3. Frequency Adjustment
- Try switching to a different frequency or channel to avoid interference from other signals.

3: Unable to Transmit

1. Check the Frequency
- Ensure you are on the correct frequency and that it is not already in use by someone else.

2. CTCSS/DCS Codes
- Verify that the CTCSS/DCS codes are correctly set. Mismatched codes can prevent successful transmission.

3. Power Settings

- Ensure the radio is set to the appropriate power level. Higher power settings can improve transmission range.

4: No Sound

1. Volume and Squelch
   - Check the volume level and adjust the squelch setting to ensure you can hear transmissions.

2. Speaker and Earpiece
   - If using an earpiece, ensure it is properly connected. If you still hear nothing, try using the radio's built-in speaker.

3. Mute Function
   - Ensure the mute function is not enabled, as this can silence all sound.

Routine Maintenance Tips

Regular maintenance is key to ensuring your Baofeng radio remains in excellent working condition. Here are some tips:

Battery Care

1. Regular Charging

   - Charge your batteries regularly, even if you're not using the radio. Avoid letting the battery completely drain, as this can reduce its lifespan.

2. Spare Batteries

   - Keep spare batteries on hand, especially if you use your radio frequently. Rotate them regularly to ensure they all remain in good condition.

Cleaning

1. Dust and Dirt

- Keep your radio clean by wiping it down with a dry, soft cloth. Avoid using water or cleaning agents that could damage the electronics.

2. Contact Points

- Clean the battery contacts and antenna connection points regularly to ensure a good connection.

Storage

1. Cool, Dry Place

- Store your radio in a cool, dry place when not in use. Avoid exposing it to extreme temperatures or moisture.

2. Protective Case

- Consider using a protective case to prevent damage if you carry your radio frequently.

Upgrading and Accessories

To get the most out of your Baofeng radio, consider investing in some useful upgrades and accessories:

Antennas

1. High-Gain Antennas
   - Upgrading to a high-gain antenna can significantly improve your radio's range and signal strength.

2. Flexible Antennas
   - Flexible antennas are more durable and less likely to break compared to standard rigid antennas.

Batteries

1. Extended-Life Batteries
   - Extended-life batteries provide longer usage times between charges, which is especially useful for extended trips or emergencies.

2. Battery Eliminators
 - These allow you to power your radio directly from a car's electrical system, providing a reliable power source during extended use.

Other Accessories

1. Speaker Microphones
 - External speaker microphones can make communication more convenient, especially if you need to keep your hands free.

2. Programming Cables
 - A programming cable allows you to easily program and update your radio's settings using software like CHIRP.

3. Protective Cases and Mounts
 - Cases and mounts can protect your radio and make it easier to carry or mount in a vehicle.

Real-Life Troubleshooting Stories

To illustrate the importance of troubleshooting and maintenance, let me share a story about a fellow radio enthusiast, Murphy. During a weekend camping trip, Murphy's Baofeng radio suddenly stopped transmitting. Instead of panicking, he calmly went through a troubleshooting checklist.

He discovered that the antenna connection was loose and that his battery needed charging. After addressing these issues, his radio was back in working order, and he was able to communicate with his group again.

Advanced Troubleshooting Techniques

For those more comfortable with technical aspects, here are some advanced troubleshooting techniques:

Software Reset

1. Factory Reset
    - If all else fails, you can perform a factory reset to restore your radio to its original settings. Be aware that this will erase all programmed frequencies and settings.

Firmware Updates

1. Check for Updates
    - Occasionally, Baofeng releases firmware updates that can improve performance or fix bugs. Check the manufacturer's website for any available updates.

2. Update Process
    - Use the programming cable and follow the instructions provided by Baofeng to update your radio's firmware.

Preparing for the Unexpected

By staying on top of maintenance and knowing how to troubleshoot common issues, you'll be well-prepared to handle any situation that arises. Remember, a well-maintained radio is a reliable radio, and knowing how to fix problems quickly can be crucial in an emergency.

You've done an excellent job mastering troubleshooting and maintenance techniques for your Baofeng radio. By keeping your equipment in top shape, you'll ensure reliable communication whenever you need it.

Remember, a well-maintained radio is not only more reliable but also more enjoyable to use. Keep practicing, stay curious, and continue refining your skills.

Chapter 7: Practical Applications of Your Baofeng Radio

Everyday Uses and Specialized Scenarios

Now that you've mastered the technical aspects of your Baofeng radio and learned how to troubleshoot and maintain it, it's time to explore the many practical applications of this versatile tool.

Whether you're using it for everyday communication, outdoor adventures, or community involvement, your Baofeng radio can be an incredibly valuable asset. Let's dive into some practical uses and specialized scenarios.

Everyday Uses

Your Baofeng radio isn't just for emergencies. Here are some ways you can incorporate it into your daily life:

Family Communication

Imagine a busy Saturday at a local festival. Cell Phone signals are weak due to the crowd, but you've got your Baofeng radio. Keeping in touch with your family becomes a breeze:

1. At Home
    - Use your radio to communicate with family members around the house or property. This can be particularly useful in large homes or properties.

2. Local Events
    - Coordinate with family or friends at crowded events like fairs, parades, or concerts. The radios help you stay connected when cell phone signals fail.

3. Shopping Trips
   - Stay in touch while shopping in large malls or separate stores. It's a great way to regroup without relying on spotty cell service.

Neighborhood Watch

Baofeng radios can enhance the effectiveness of neighborhood watch programs. Here's how:

1. Routine Patrols
   - Use radios to coordinate patrols and report suspicious activity. This ensures quick communication and response.

2. Emergency Coordination
      - Quickly alert neighbors about emergencies or safety concerns. Radios can be more reliable than phones during power outages.

Hobbies and Recreation

Your Baofeng radio can also enhance your hobbies:

1. Model Rocketry
   - Communicate with team members during launches. Coordinate safety checks and recovery operations.

2. Cycling or Running Groups
   - Stay connected with your group during long rides or runs. Radios ensure everyone is accounted for and can share updates or emergencies.

Outdoor Adventures

Baofeng radios are perfect companions for outdoor enthusiasts. Here's how they can be useful:

Hiking and Camping

1. Trail Communication

- Stay in touch with your group while hiking. Share updates on trail conditions, wildlife sightings, and more.

2. Camp Coordination
  - Use radios to coordinate activities and ensure everyone's safety in the campsite.

Off-Roading and Overlanding

1. Vehicle Convoys
  - Coordinate with other vehicles in your convoy. Share information about road conditions, obstacles, and navigation.

2. Emergency Situations
  - Quickly call for help if you encounter trouble. Radios can be lifesavers in remote areas without cell coverage.

Boating and Fishing

1. Marine Communication
   - Communicate with other boats or shore bases. Share information about fishing spots, weather updates, and emergencies.

2. Safety Coordination
   - Use radios to coordinate safety measures and ensure everyone on board is accounted for.

Community Involvement

Baofeng radios can play a significant role in community activities and volunteer efforts:

Event Coordination

1. Festivals and Fairs
   - Coordinate with volunteers and staff to manage large events efficiently.

Radios help ensure smooth communication and quick responses to issues.

2. Parades and Sports Events
- Use radios to coordinate logistics, manage crowds, and communicate with security personnel.

Disaster Response Teams

1. Local Preparedness
- Join or form a local disaster response team. Use radios to coordinate efforts, share information, and respond quickly to emergencies.

2. Training Drills
- Participate in training drills to practice using radios in simulated emergency scenarios. This prepares you for real-life situations.

Specialized Uses

Your Baofeng radio can be adapted for specialized scenarios, enhancing its versatility:

Amateur Radio (Ham) Activities

1. Ham Radio Events
    - Participate in amateur radio events and contests. Use your Baofeng radio to communicate with other operators and practice your skills.

2. Learning and Experimentation
    - Explore different frequencies and modes. Experiment with antenna setups, power settings, and more to enhance your understanding and skills.

Search and Rescue (SAR)

1. SAR Operations
    - Use your radio to coordinate search and rescue operations.

Clear communication is vital for the safety and success of these missions.

2. Training and Drills
 - Participate in SAR training sessions to practice using your radio in simulated rescue scenarios.

Farming and Agriculture

1. Farm Coordination
 - Use radios to communicate with workers across large farms. Coordinate tasks, share updates, and ensure safety.

2. Livestock Management
 - Communicate quickly during livestock management activities. Radios help keep everyone informed and coordinated.

Enhancing Your Skills

To make the most of your Baofeng radio, continuously enhance your skills:

1. Practice Regularly
   - Use your radio frequently to become more comfortable with its features and functions.

2. Join Clubs and Groups
   - Join local radio clubs or online groups. These communities offer valuable resources, support, and opportunities to practice.

3. Stay Informed
   - Keep up with the latest developments in radio technology and regulations. This ensures you're always using your radio effectively and legally.

Optimizing Your Radio Setup

To ensure your radio setup is always optimized for the best performance, consider the following tips:

Regular Updates and Check-Ups

1. Firmware Updates

   - Check for firmware updates regularly and install them to ensure your radio has the latest features and bug fixes.

2. Routine Maintenance

   - Perform routine maintenance checks on your radio, antenna, and batteries. Keeping your equipment in good condition ensures reliable performance.

Customizing Settings

1. Personalized Settings
 - Customize your radio settings to match your preferences and needs. This includes adjusting power levels, setting up scan lists, and organizing memory channels.

2. Shortcut Keys
 - Use shortcut keys and programmable buttons to access frequently used features quickly. This can save time and enhance your radio's usability.

You've now explored the various practical applications of your Baofeng radio. From outdoor adventures and community involvement to specialized uses, these radios prove to be indispensable tools.

Remember, the key to mastering your Baofeng radio is continuous learning and practice. Stay curious, keep exploring, and don't be afraid to try new things. The future of radio communication is bright, and I'm excited to see how you'll be part of it.

Keep your radio close, your mind open, and your spirit adventurous. You never know when the next great communication breakthrough will happen—or when you'll be the one to discover it.

Chapter 8: The Future of Baofeng Radios

Embracing Tomorrow's Technology Today

By now, you've learned the ins and outs of using your Baofeng radio, and you've seen just how versatile and powerful these devices can be.

Emerging Trends in Radio Communication

Technology is always evolving, and the world of radio communication is no exception. Here are some trends and advancements that might shape the future of Baofeng radios:

Digital Communication Modes

1. DMR (Digital Mobile Radio)

- DMR is already making waves in the radio community.
It offers clearer audio, better range, and more efficient use of frequencies. Imagine chatting with crystal-clear sound, even in noisy environments.

2. Fusion and D-STAR

- These digital modes are becoming more popular among amateur radio operators. They provide robust features like GPS integration, text messaging, and internet connectivity, making communication more versatile and fun.

Software-Defined Radios (SDR)

1. Flexibility and Customization

- SDR technology allows radios to be more flexible and customizable. With SDR, you can update your radio's features and capabilities through software, keeping your device up-to-date with the latest advancements without needing to buy new hardware.

2. Broad Frequency Range

- SDRs can cover a broader frequency range than traditional radios, giving you access to more bands and modes. This opens up new possibilities for experimentation and exploration.

Integration with Smartphones and IoT

1. Smartphone Apps

- Imagine controlling your Baofeng radio with a smartphone app. Some modern radios already offer this functionality, allowing you to program frequencies, adjust settings, and even communicate directly from your phone.

2. Internet of Things (IoT)

- The IoT is connecting devices like never before.
Future Baofeng radios might integrate with smart home systems, wearable tech, and other IoT devices, making communication more seamless and integrated into your daily life.

Preparing for Tomorrow

To stay ahead of the curve and make the most of these advancements, here are some tips:

Continuous Learning

1. Stay Informed

- Keep up with the latest news and developments in the radio community. Websites, forums, and magazines are great resources to stay informed about new technologies and trends.

2. Join Clubs and Groups

- Connect with other radio enthusiasts. Clubs and online groups provide a wealth of knowledge, support, and opportunities to learn and grow.

Experiment and Innovate

1. Try New Modes

- Don't be afraid to experiment with new digital modes like DMR or Fusion. You'll not only learn new skills but also expand your communication capabilities.

2. DIY Projects

- Get hands-on with DIY projects. Building your own antennas, setting up an SDR, or integrating your radio with a smartphone app can be both fun and educational.

Invest in Future-Proof Equipment

1. Upgradable Radios

   - Consider investing in radios that offer firmware updates and software-defined capabilities. This ensures your device can adapt to new technologies without becoming obsolete.

2. High-Quality Accessories

   - High-quality antennas, batteries, and other accessories can enhance your radio's performance and longevity, making it ready for future advancements.

The key to mastering any technology is continuous learning and experimentation. Stay curious, keep exploring, and don't be afraid to try new things. The future of radio communication is bright, and I'm excited to see how you'll be part of it.

Keep your radio close, your mind open, and your spirit adventurous. You never know when the next great communication breakthrough will happen—or when you'll be the one to discover it.

www.ingramcontent.com/pod-product-compliance
Lightning Source LLC
Chambersburg PA
CBHW071935210526
45479CB00002B/696